Title: Organic-Way Mathematics: The Essence of Problem Solving - Grades 5-6-7 - Student Workbook
Author: Archangelo Joseph, Ed.D.
Layout and Cover: Kate Keverline

Copyright © 2023, Bridgevision Production LLC.

All rights reserved. This book or any portion thereof may not be reproduced or used in any manner whatsoever without the express written permission of the publisher except for the use of brief quotations in a book review. For permission requests, contact the publisher:

Bridgevision Production LLC
PO BOX 9767
Washington, DC 20016
erilandesully@gmail.com
www.bridgevisionllc.com

Library of Congress Catalogue-in-Publication Data is available upon request. Originally published: 2023.

ISBN: 978-1737678991

ORGANIC-WAY MATHEMATICS: THE ESSENCE OF PROBLEM SOLVING
STUDENT WORKBOOK 5-6-7
Copyright © 2023

FIVE STAGES OF OMEPS

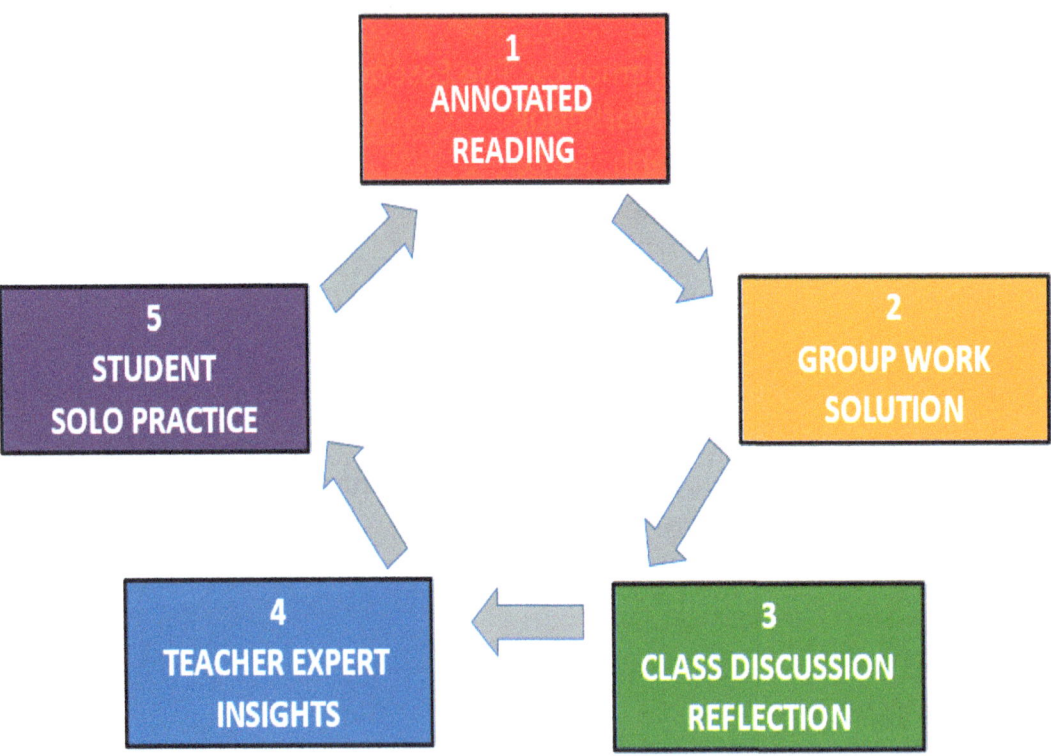

ORGANIC-WAY MATHEMATICS: THE ESSENCE OF PROBLEM SOLVING
STUDENT WORKBOOK 5-6-7
Copyright © 2023

PROBLEM-1
ORGAMATH TIME

GROUP TASK-1

The 90-minute Organic-way Math lesson allocates time to its components, as follows:

1. Annotated Reading: 1/18 of the total time.
2. Group Solution   The triple of the Annotated Reading
3. Class Discussion: 2,700 seconds
4. Teacher Insights   ?
5. Solo Practice   0.25 hour

- Determine the amount of time (in minutes) left for Teacher Insights.
- Complete the OrgaMath Time Table.
- Complete the ORGAMATH Time Circle Graph.
- Analyze the data and write your observation.

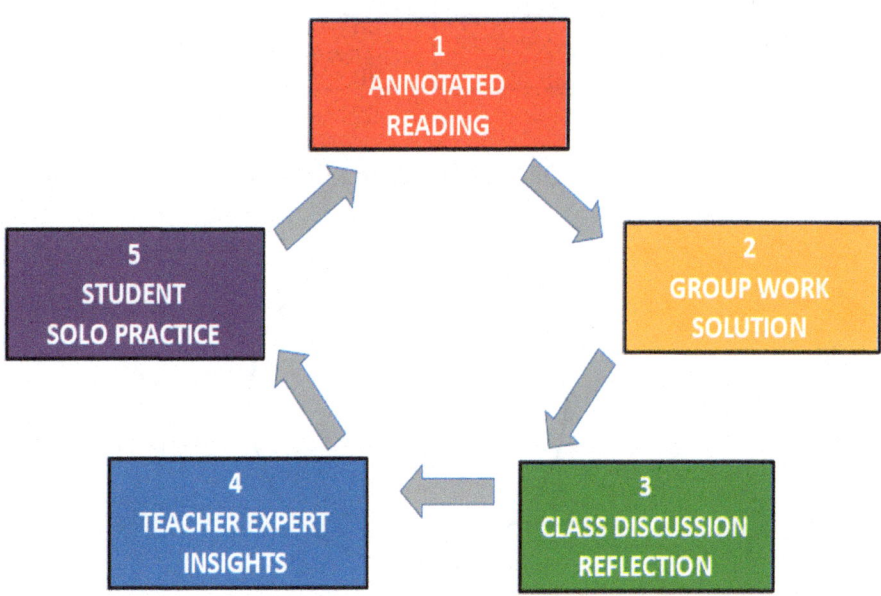

SOLO PRACTICE-1

Imagine that the allotted time for the entire Organic-way Math lesson were one hour. How much time would be proportionally devoted to each lesson stage of the lesson. Complete the relevant table and circle graph.

OrgaMath Time Table				
OMEPS Stage	Allotted Time	Fraction	Decimal	Percentage
Annotated Reading	5 min.			
Group Solution				
Class Discussion				
Teacher Insights				
Solo Practice				
Whole Lesson		$\frac{90}{90} = 1$	1.00	100%

ORGAMATH Time Circle Graph

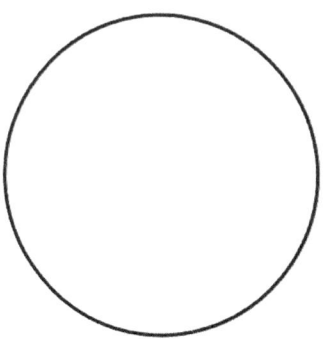

PROBLEM-2
STUDENT-TEACHER RATIO

GROUP TASK-2

A lower-grade Co-teaching class requires a student-teacher ratio of 5:1. Two-thirds of the students are boys and the rest are girls. The class has 15 students.
- Determine the number of students of each gender.
- Including teachers, complete the Table Student-Teacher Ratio.
- Complete the Circle Graph for Student-Ratio.

Table for Student-Teacher Ratio				
People	Number	Fraction	Decimal	Percentage
Boys		$\frac{10}{18} = \frac{5}{9}$		
Girls				
Teachers				
Total				100%

Circle Graph for Student-Teacher Ratio

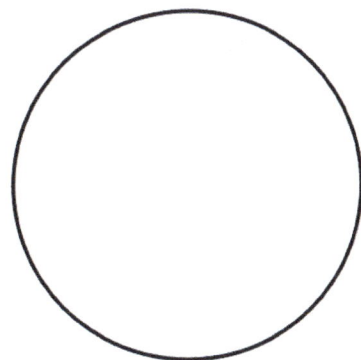

ORGANIC-WAY MATHEMATICS: THE ESSENCE OF PROBLEM SOLVING
STUDENT WORKBOOK 5-6-7
Copyright © 2023

SOLO PRACTICE-2

- Imagine that the same school had 200 students, and all the classes are co-teaching with the same ratio 5:1 of teachers to students. Determine the number of students of each gender.
- Including teachers, complete the Table Student-Teacher Ratio.
- Complete the Circle Graph for Student-Ratio.

People	Table for Student-Teacher Ratio			
	Number	Fraction	Decimal	Percentage
Boys				
Girls				
Teachers				
Total	240		1.00	

Circle Graph for Student-Teacher Ratio

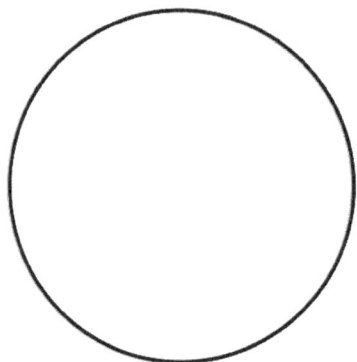

PROBLEM-3
MIXED BEANS

GROUP TASK-3

1. Explore your batch of mixed beans.
2. Estimate the total number to be compare later with the actual number.
3. Group them by type or color.
4. Count the number pf bean in each group.
5. Complete the Table for Mixed Beans below and Circle Graph below, assuming that each bean is worth $0.15. Use as many rows as you need.
6. Write your observation.

Table for Mixed Beans					
Color	Quantity	Fraction	Decimal	Percentage	Cost
Black	10				
Green	3				
Pinto	2				
Red	5				
Total	20				

Observation:

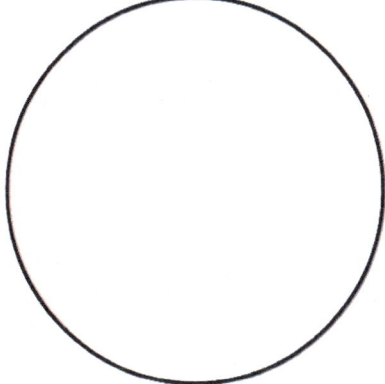

Circle Graph for Mixed Beans

SOLO PRACTICE-3

Without looking, scoop a smaller number of mixed beans to reproduce on a smaller scale a Solo task similar to Group Task-1.

ORGANIC-WAY MATHEMATICS: THE ESSENCE OF PROBLEM SOLVING
STUDENT WORKBOOK 5-6-7
Copyright © 2023

PROBLEM-4
SUPERMARKET SHOPPING

GROUP TASK-4

Based on the accompanying food flyers go to the market and purchase the food items below. Round all costs to the nearest dollar.
Complete the Table for Supermarket Shopping.
Complete the Circle Graph for Supermarket Shopping.
Write your observation.
- 4 packs of sweet corn
- 3 pints of strawberry
- 5 bags of Cello spinach
- 10 lbs. of chicken drumsticks

Note: Round all costs to the nearest cent.

ORGANIC-WAY MATHEMATICS: THE ESSENCE OF PROBLEM SOLVING
STUDENT WORKBOOK 5-6-7
Copyright © 2023

ORGANIC-WAY MATHEMATICS: THE ESSENCE OF PROBLEM SOLVING
STUDENT WORKBOOK 5-6-7
Copyright © 2023

GROUP TASK-4 (continued)

Picture	Quantity	Cost	Fraction	Decimal	Percentage
	4 packs	4 × 1.89 = $7.56	$\dfrac{7.56}{23.08}$		33%
				0.167	
Total Budget	xxxxxxx	$23.08	$\dfrac{23.08}{23.08}$		100%

Observation Report:

Supermarket Table Circle Graph

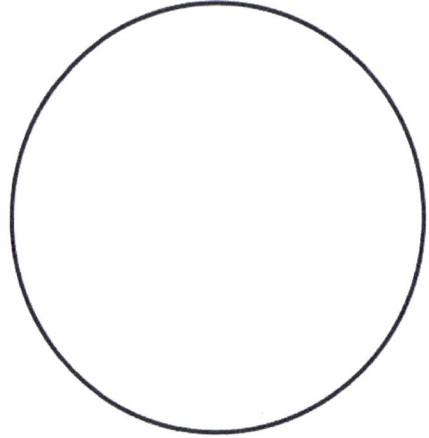

SOLO PRACTICE-4

1. Use only $10.00 from your wallet to purchase the most quantity of items. Indicate the percent of the 10-dollar budget allocated to each category.

2. A senior citizen purchases 8 lbs. of Sirloin Tip Steak at 10% discount. Determine her change a 50-dollar bill.

PROBLEM-5
BLACK FRIDAY

GROUP TASK-5

On Black Friday, Linda went to Flatbush Avenue in Brooklyn. There, she purchased on sale three sweaters: one black, one blue-grey, and one white-grey. The black sweater cost $20.75. The blue-grey was $1.25 less than the black. The white-grey sweater was as much as the black sweater and the blue-grey sweater combined.

The tax is 8.75%.

Determine the change (if any) from a 100-dollar bill.

SOLO PRACTICE-5

1. Sue purchased one pair of pants for $50.99, and one blouse that cost $23.60 less than the pair of pants. She received $21.62 as change. How much did Sue tender to the cashier?

2. Louis bought one shirt that cost as much as 3 pairs of socks combined. The price of one pair of socks was $18.45. The tax was 8.75%. How much did he owe to the cashier?

PROBLEM-6
TAXI RIDE

GROUP TASK-6

Last Sunday at 7:00 p.m., Denise hailed a taxi cab for a 10-mile ride from Brooklyn to Manhattan. Inside the vehicle, the fare table below indicates the charges. In addition, Denise left a 15% tip for the driver. Determine the amount she spent in all on the ride.

FARE TABLE	
Initial charge	$3.50
Each 1/5 mile	$0.75
Every 2-minute stop	$0.40
Weekday surcharge (8 pm – 6 am)	$1.00
Weekend surcharge (4 pm – 8 pm)	$0.50

SOLO PRACTICE-6

One weekday around 5:00 am, Jazzy took the same taxi from Queens to The Bronx for a 6-mile ride. At destination, she tipped the driver with a 5-dollar bill. Determine the amount Jazzy has spent in all on the 6-mile ride.

ORGANIC-WAY MATHEMATICS: THE ESSENCE OF PROBLEM SOLVING
STUDENT WORKBOOK 5-6-7
Copyright © 2023

PROBLEM-7
BERMUDA FISH

GROUP TASK-7

A fisherman caught a giant 60-kg red snapper off the coast of Bermuda. He cut the fish into 4 sections. He sold Mid-Section B for $40, which was 5 times more than Mid-Section A. The Head-Section weighed 1/3 of the fish.
Complete the Bermuda Fish Table and the Bermuda Fish Circle Graph.
Write what you have observe in the table.

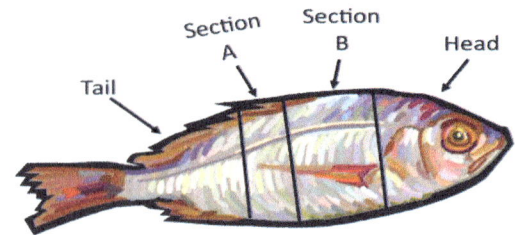

Section	Weight	Fish Fraction	Notation in Decimal	Percentage	Cost
Head Section	20 kg				
Mid-Sect. A					
Mid-Sect. B					$40.00
Tail Section					
Whole Fish	60 kg	$\frac{60}{60} = 1$		100%	

Bermuda Fish Circle Graph

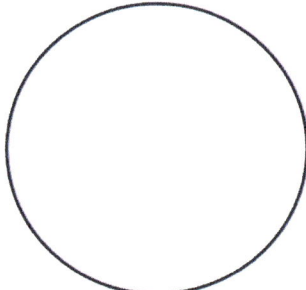

SOLO PRACTICE-7

If the total weight of the fish were 45 pounds, complete the table according to the proportions found in the Group Task Problem-5. The unit cost remains constant ($1.6 per lb.

Bermuda Fish Table

Section	Weight	Fish Fraction	Notation in Decimal	Percentage	Cost
Head Section				33%	
Mid-Sect. A					
Mid-Sect. B	18.90 lbs.				$30.24
Tail Section					
Whole Fish	45 lbs.	$\frac{45}{45} = 1$		100%	

Bermuda Fish Circle Graph

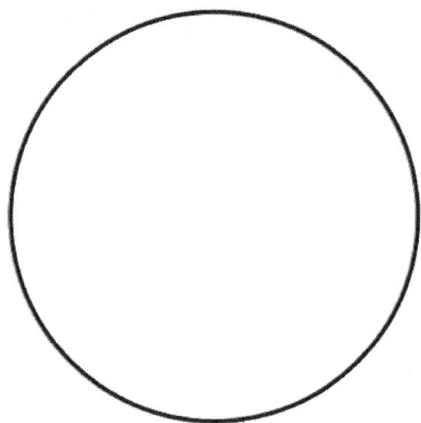

PROBLEM-8
OPEN BOX

GROUP TASK-8

1. Given a regular sheet of paper or origami/oaktag, find the dimensions in cm (length and width). Then find and record the area.
2. Follow attentively the teacher's instructions to construct an open box or rectangular prism (see series of images below).
3. Determine the surface (total) area of the five faces of the open box, and round all areas to the nearest ten.
4. Compare this total area with the area of the original sheet of paper.
5. Complete the Open Box Table, the Front Face $19.25 to be painted.
6. Complete the Open Boc Circle Graph.
7. What have you observed?

ORGANIC-WAY MATHEMATICS: THE ESSENCE OF PROBLEM SOLVING
STUDENT WORKBOOK 5-6-7
Copyright © 2023

Folding the Sheet of Paper into an Open Box

A. Hotdog Fold

B. Hamburger Fold

C. Prying opening Hamburger Fold

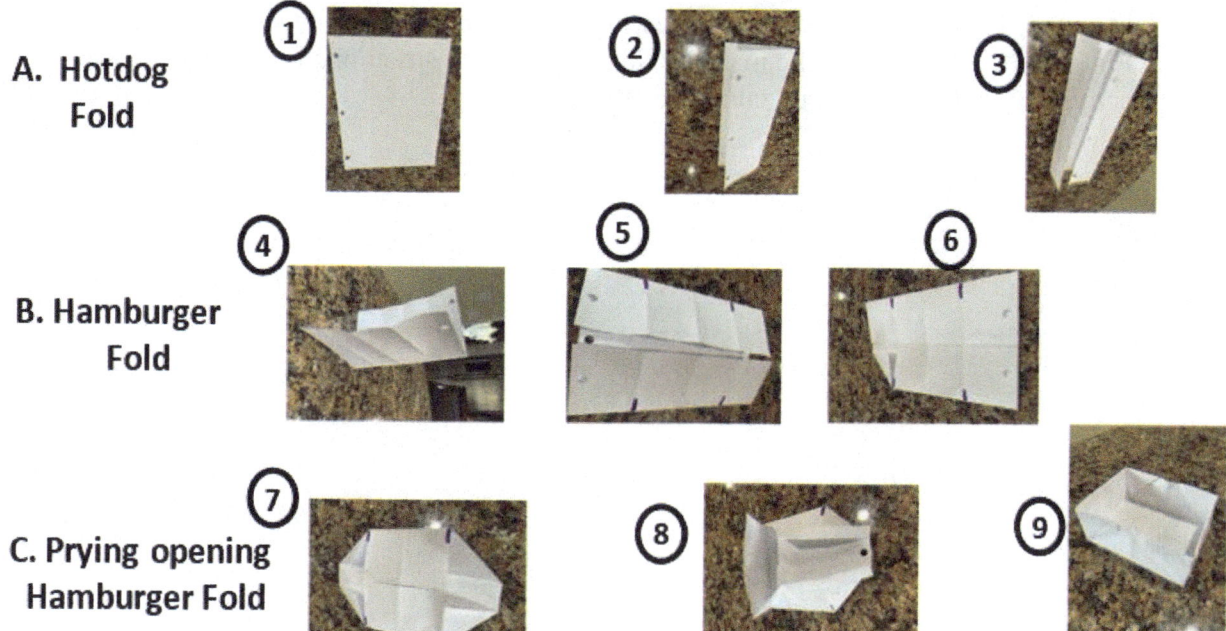

GROUP TASK-8 (continued)

Open Box Table						
Prism Face	Area	Fraction	Decimal	Percentage	Cost
Front					$19.25
Back					
Right					
Left					
Bottom					
Hidden Area					
Total Area (Orig. Paper)					

21

SOLO PRACTICE-8

Use an origami or oaktag sheet of paper of a size differing from the sheet used in Problem-6A Group. Fold the material similarly to the instructions given in the group activity. Construct a relevant table and circle graph.

PROBLEM-9
PRISM-CYLINDER

GROUP TASK-9

Revisit Prolem-6 (Open Box).
Reconstruct the box using origami paper or oaktag. Then construction a cylinder of your own dimensions that would have the same amount of cereal (capacity) as the prism (open box).
Justify your result (equal capacity) by pouring cereal in both containers.

SOLO PRACTICE-9

Construct a task similar to the previous Group Task of Problem 6B along with the a proposed solution.

ORGANIC-WAY MATHEMATICS: THE ESSENCE OF PROBLEM SOLVING
STUDENT WORKBOOK 5-6-7
Copyright © 2023

PROBLEM-10
LIBRARY PROPOSAL

GROUP TASK-10

The floor of a local library is to be renovated according to the sketch below.
Carefully examine the data on the sketch.
Complete the Table for Library Proposal.
Complete the Circle Graph for Library Proposal.
Write your observation.

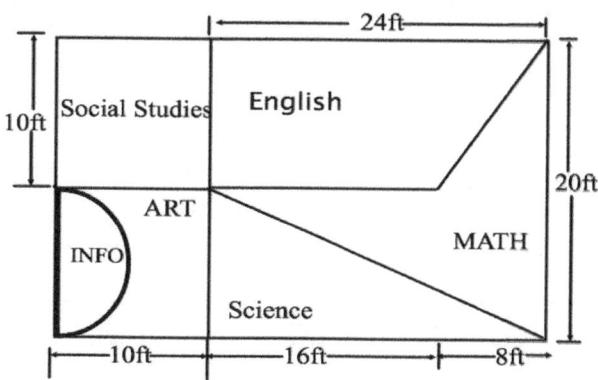

Table for Library Proposal					
Lib. Section	Area	Fraction	Decimal	Percentage	Cost
Social Studies					
Science					$720
English		$\frac{200}{680} = \frac{5}{17}$			
Info					
Art					
Math					
Total Area	680 ft²			100%	

25

ORGANIC-WAY MATHEMATICS: THE ESSENCE OF PROBLEM SOLVING
STUDENT WORKBOOK 5-6-7
Copyright © 2023

SOLO PRACTICE-10

Below is the blueprint of a City Shelter Building in case of emergency. Each room will be built at $500 per square yards. Calculate the total cost. Complete the table.

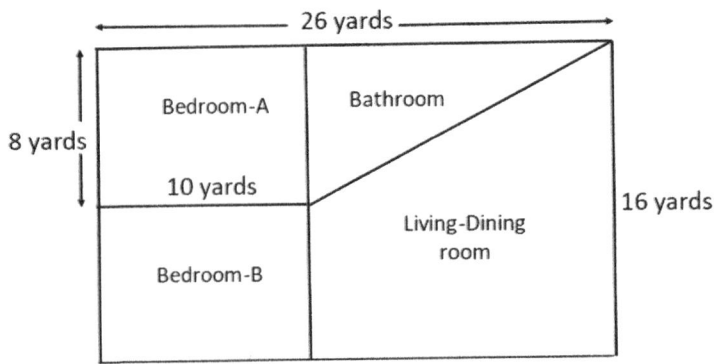

City Shelter Building Table					
Room	Area	Fraction	Decimal	Percentage	Cost
Bedroom A					
Total Area					$257,291.84

PROBLEM-11
PRIME & COMPOSITE NUMBERS

GROUP TASK-11

Complete the 100-natural number chart below. Examine the chart and detect some patterns. Define prime numbers and color them red on the chart. Complete the Prime-and-Composite Numbers Table below. What have you observed?

1									
									100

Prime-and-Composite Numbers Table

Natural Numbers	Quantity	Fraction	Decimal	Percentage
Prime	25			
Composite				
None				
Total	100	$\frac{100}{100} = 1$	1.0	100%

Prime-and-Composite Circle Graph

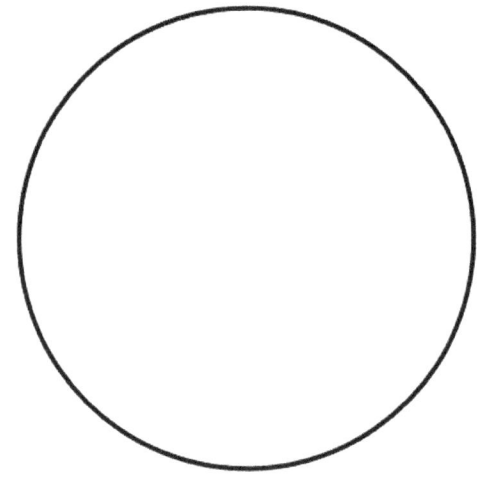

SOLO PRACTICE-11

Write the set N of all natural numbers, such that $100 < N < 121$.
Then circle all prime numbers.
Indicate the percentage of:
- Prime numbers
- Composite numbers
- None.

PROBLEM-12
GOOSE PLAY

GROUP TASK-12

One summer day, 630 brown geese and 370 white geese were found respectively on the West Bank and East Bank of the Huson River in Nyack. After a while, $\frac{2}{5}$ of the white geese swam to the west bank, and $\frac{3}{7}$ of the brown geese paddled to the east bank. Then 50% of the remaining white geese went westward, while 60% of remaining brown geese traveled eastward.

Determine the number of each kind of geese on each bank of the Huson River by completing the accompanying table.

Examine the table and write your observation.

West Bank			East Bank	
Geese White	Geese Brown		Geese White	Geese Brown
0	630	Hudson River		

SOLO PRACTICE-12

On the left shore of a river, there were 600 black geese and 400 grey geese. After 30 seconds, $\frac{1}{2}$ of the black geese and $\frac{1}{2}$ grey geese swam to the right shore. Then from the leftover, 175 black geese and $\frac{1}{2}$ of the grey geese also crossed to the left shore. Finally, 75 black geese and 160 grey geese returned to the right shore of the river.

Complete the Crazy Goose Play Table.
Record the number of geese in each column after each trip.
Write about what you have observation the table.

Left Shore			Right Shore	
Black Geese	Grey Geese		Black Geese	Grey Geese
600	400	River		

1. On a river shore, the number of geese was increased to 1,000 after 570 geese had arrived to join the flock. How many geese were there at the beginning?

PROBLEM-13
EXPERIMENTAL-&-THEORETICAL PROBABILITY

GROUP TASK-13

Follow the teacher's instructions to conduct and compare an experimental probability and a theoretical probability, as follows:

1. First, conduct an experiment by tossing, at least 50 times, two dice of different colors and record the sums of numbers (outcomes) on a tally.

   Calculate the probability of each experimental outcome.

2. Second, complete the accompanying Addition Table and the related Sum-Probability Table. What have you observed?

Addition Table

+	1	2	3	4	5	6
1						
2						
3						
4						
5						
6						

Prob.	Sum Two	Sum Three	Sum Four	Sum Five	Sum Six	Sum Seven	Sum Eight	Sum Nine	Sum Ten	Sum Eleven	Sum Twelve
Sum-Probability Table											
Fraction											
Decimal											
Percent											

SOLO PRACTICE-13

1. Determine the probability for a blue marble to be picked at random (without looking) from a sack that contains 4 white marbles and 6 yellow marbles.

2. A sack contains 4 white marbles and 6 yellow marbles.
   a). Find the probability for a yellow marble be picked at random on the first try.
   b). Determine the probability that a white marble be picked at random on the second try without replacement.

3. A sack contains 5 red marbles, 7 blue marbles, and 8 green marbles.
   a). Find the probability that a blue marble be picked at random on the first try.
   b). Determine the probability that a green marble be picked at random on the second try with replacement second try.

ORGANIC-WAY MATHEMATICS: THE ESSENCE OF PROBLEM SOLVING
STUDENT WORKBOOK 5-6-7
Copyright © 2023

PROBLEM-14
STUDENT-WORK ANALYSIS

GROUP TASK-14

Analyze the solution composed by the student (Paul Andre) to the problem below based on the accompanying rubrics. Indicate the score you would give Andre's work.

(Problem whose solution is to be analyzed):

A professional earns a gross pay of $6,500. From that amount, $\frac{1}{5}$ is withheld for Federal taxes and 8% for local taxes. In addition, $200 and $150 are deducted respectively for future Medicare and Social Security benefits. After all deductions, the check indicates a net pay of $ 4,330. Investigate whether the net pay is correct.

Student: Paul Andre   Class: 602   Date: 02-28-22

Solution

Federal taxes: $\frac{1}{5} \times \$6,500 = \$1,300$

Local taxes: $\$6,500 \times 0.08 = \$510$

Total deductions:

$\$1,300 + \$510 + \$200 + \$150 = \$2,160$

Net pay: $\$6,500 - \$2,160 = \$4,340$

Answer: No. The paycheck is not correct. It misses $10.

Performance Levels	Rubric Performance Descriptors
Level 4	An appropriate strategy is used. For example, the student knows how to take 1/5 of $6,500 and when to perform such basic operations as multiplication, addition, and subtraction. A solution line / phrase or label is provided for each operation. The process is coherent, and there is no mistake.
Level 3	An appropriate strategy is used. There may be one minor error at the beginning, with a consistent logic. The entire process is good, but the final answer may be wrong due to a computational error previously made. Some solution lines are missing.
Level 2	A good strategy was attempted. There are some errors of conceptual understanding leading to a wrong answer. Part of process may be good. Solution lines are missing.
Level 1	An inappropriate strategy is used. There multiple major errors some of which are of conceptual understanding and procedural fluency. Solution lines are missing or confusing.

SOLO PRACTICE-14

Compose a word problem and a related solution. Also, develop a set of rubrics that will enable the student-teacher to score the solution you have composed.

ORGANIC-WAY MATHEMATICS: THE ESSENCE OF PROBLEM SOLVING
STUDENT WORKBOOK 5-6-7
Copyright © 2023

PROBLEM-15
PATTERN BLOCKS

GROUP TASK-15

Explore your set of pattern blocks. Examine and identify the shape of each piece. Establish the relationships among them, and particularly the relationship (fraction, decimal and percentage) of each smaller polygon to the hexagon.
Complete the table proportionally to the cost ($48) of the trapezoid. What have you observed?

Shape	Polygon	Fraction	Decimal	Percentage	Cost
(trapezoid)					$43.05
(hexagon)					

Shape	Polygon	Fraction	Decimal	Percentage	Cost
(triangle)					
(hexagon)					

Shape	Polygon	Fraction	Decimal	Percentage	Cost
◇					
⬡					

Shape	Polygon	Fraction	Decimal	Percentage	Cost
☐					
⬡					

Observation:

SOLO PRACTICE-15

Explore again your set of pattern blocks. Examine and identify the shape of each piece. Establish the relationships among them, and particularly the relationship (fraction, decimal and percentage) of each polygon to the rhombus, costing $60. Construct a series of tables similar to those on Problem-12 - Pattern Blocks-Group Task. What have you observed?

PROBLEM-16
COORDINATED POLYGONS

GROUP TASK-16

1. On the grid below, plot the points: A(3,3); B(5,3); E(4,0); F(2,0) Connect the points to get Polygon ABEF.
2. Plot D(6,0). Connect B, D, and E.
3. Plot C(6,6). Connect the points to get Polygon BCD.
4. Find the areas Polygons ABEF, BDE, BCD, and ABCDF.
5. Complete the table and the pie graph below, using the areas of the polygons.
6. Write your observation.

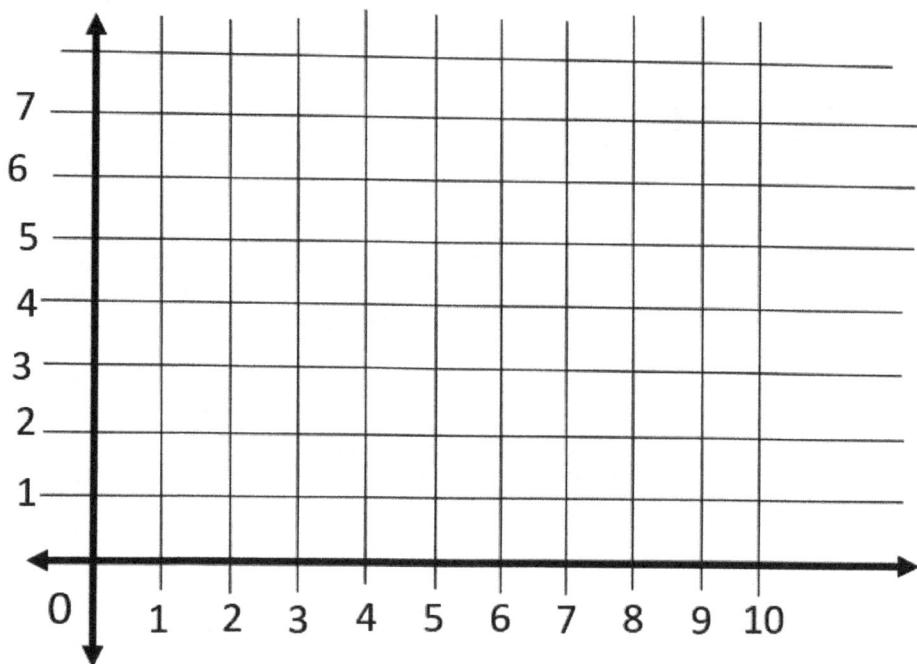

Coordinated Polygons Table						
Polygon	Area	Fraction	Decimal	Percentage	Cost
ABEF					
BDE					
BCD					$425.50
ABCDF	12 u²			100%	

Coordinated Polygons Circle Graph

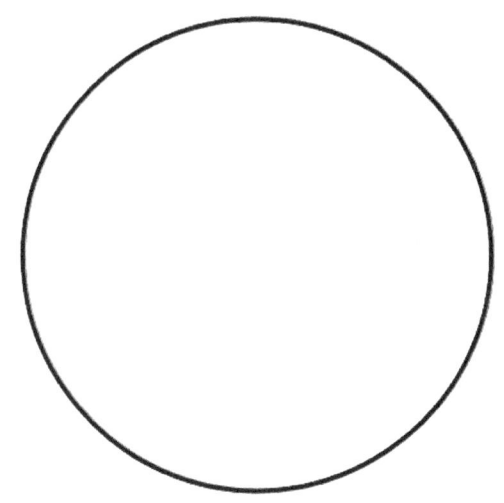

SOLO PRACTICE-16

1. On the grid below, plot the points: A(3,3); B(5,3); F(4,0); G(2,0) Connect the points to get Polygon ABFG. Find the area.
2. Plot E(6,0). Connect B and E; then E and F. Find area of Polygon BEF.
3. Plot C(5,7) and D(6,7) Connect the points to get Polygon BCDE. Find the area.
4. Complete the table and the circle graph.
5. Write your observation.

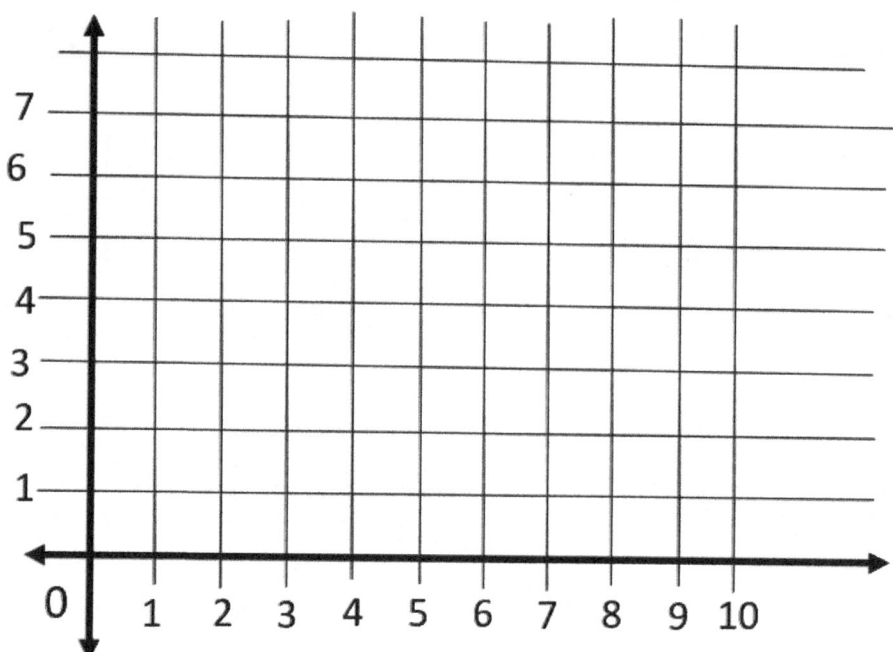

Coordinated Polygons Table						
Polygon	Area	Fraction	Decimal	Percentage	Cost
ABFG					
BEF					$81
BCDE					
ABCDEG	14.5 u²				

Coordinated Polygons Circle Graph

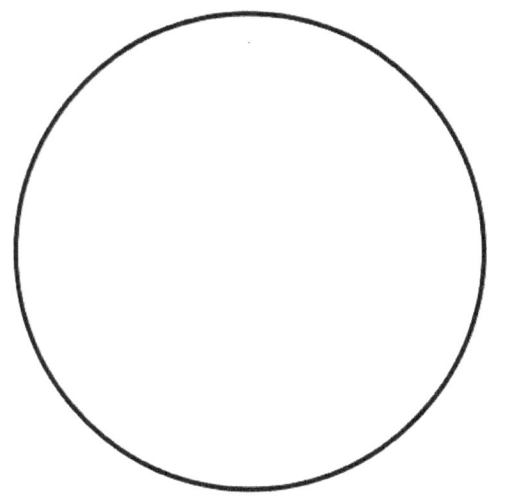

ORGANIC-WAY MATHEMATICS: THE ESSENCE OF PROBLEM SOLVING
STUDENT WORKBOOK 5-6-7
Copyright © 2023

PROBLEM-17
WHAT'S π?

GROUP TASK-17

Conduct the following experiment including your findings as to the meaning of π..

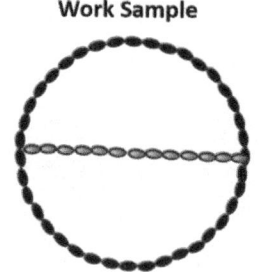
Work Sample

1. Use some household objects to trace four circles A, B, C, and D of different sizes.
2. Trace the diameter of each circle.
3. Select dried beans of the same size and glue them on the circumference of the circle. Record that number of beans.
4. Glue beans similarly on the diameter of each circle, and record that number.
5. Use these new-found data to compete the What's Pi Table.
6. Analyze the table and report your findings.

What's Pi Table?			
Circles	Circumference	Diameter	Circumference to Diameter Ratio
Circle A			
Circle B			
Circle C			
Circle D			

SOLO PRACTICE-17

Repeat a similar experiment with as many circles as you wish to prove the value of π.

ORGANIC-WAY MATHEMATICS: THE ESSENCE OF PROBLEM SOLVING
STUDENT WORKBOOK 5-6-7
Copyright © 2023

PROBLEM-18
MAGIC TANGRAM

GROUP TASK-18

Using the seven pieces of your Tangram set to complete the task below.

1. Explore your Tangram set.
2. Tessellate the pieces into a larger square. Explain how you have arrived at doing it.
3. Determine the area of each polygon to the nearest mm^2, using many strategies.
4. Complete the Magical Tangram Table and the Magic Tangram Circle Graph.
5. Write your observation.

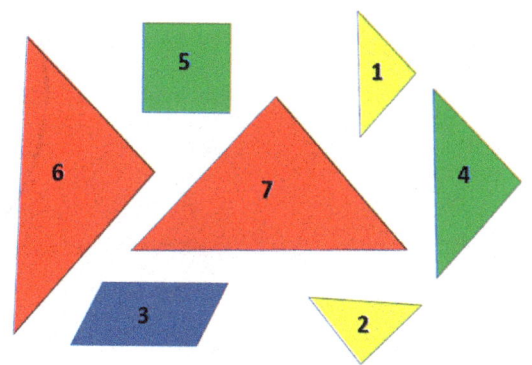

Magic Tangram Table						
Polygon	Name	Area	Fraction	Decimal	Percentage	Cost
1	Small Triangle					
2	Small Triangle					
3	Parallelogram			0.125		
4	Medium Triangle					$437.50
5	Square					
6	Large Triangle					
7	Large Triangle					
Total	xxxxxxxxxx			1.0		

Magic Tangram Circle Graph

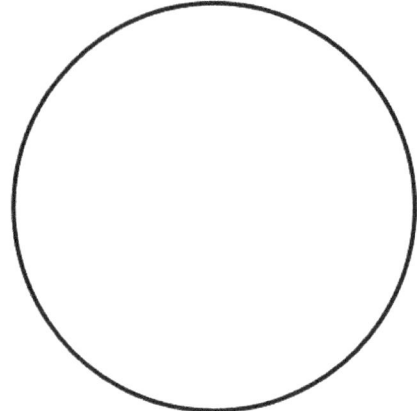

Observation:

SOLO PRACTICE-18

From the original 7 pieces of the Tangram Select any set of five pieces of the Tangram set, remove the two largest triangles. Tesselate the rest of the polygons into a jumbo triangle. and construct a table and circle graph similar to the ones in the previous Group Task. Use $40 as the cost of the square.

What have you observed in comparison to the previous table and circle graph constructed during the previous Group Task.

Magic Tangram Table						
Polygon	Name	Area	Fraction	Decimal	Percentage	Cost
1	Small Triangle					
2	Small Triangle					
3	Parallelogram					
4	Medium Triangle					
5	Square					$40
Total	xxxxxxxxxx					

Magic Tangram Circle Graph

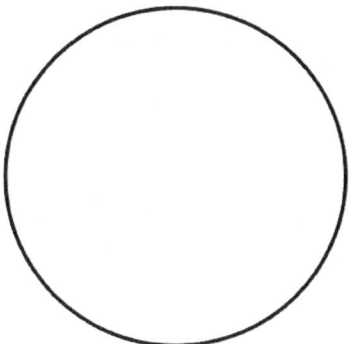

ORGANIC-WAY MATHEMATICS: THE ESSENCE OF PROBLEM SOLVING
STUDENT WORKBOOK 5-6-7
Copyright © 2023

PROBLEM-19
BEST SHOOTER

GROUP TASK-19

One summer, five youngsters Rich, Jazz, Rick, Shy, and Fah formed a household basketball team in Nyack. Before confronting a neighborhood team, they decided to find out who among them was the best shooter. To do so, they agreed to a shootout competition, yielding the following results in shot made and shots attempted.

1. Determine the best shooter while completing the Best Shooter Table
2. Rank all the shooters.
3. Find the team average.

Best Shooter Table					
Shooters	Shots Made	Shots Attempted	Fraction	Decimal	Percentage
Rich					
Jaz					
Rick					
Shy					
Fah					

SOLO PRACTICE-19

Examine the score table below. Based on your analysis, determine who is the most prolific shooter.

\multicolumn{6}{c}{Best Shooter Table}
Shooters	Shots Made	Shots Attempted	Fraction	Decimal	Percentage
Ann	12	16			
Max	10	14			
Sue	10	12			
Tom	9	16			
Fey	11	13			

PROBLEM-20
MANGO JUICE

GROUP TASK-20

Narrator: Ladies and gentlemen, please welcome Jeanne and Mary. Jeanne is a young lady who loves mango juice. Mary is a peddler who wants to sell as many mangoes as possible. This morning out goes Jeanne in search of her favorite fruit. As Jeanne crosses a street corner, a soft voice breaks the silence in her mind.

Mary:	My dear beautiful lady, please come and check my mangoes.
Jeanne:	Okay. No problem. Indeed, I need some mangoes to make juice.
Mary:	That's a good idea. Today they're fresh and delicious.
Jeanne:	How do you sell these Francique mangoes?
Mary:	Three for five dollars.
Jeanne:	Okay. No problem. I'll take eight.
Narrator:	Freeze! How much does Jeanne owe for the eight mangoes?
Mary:	Mmm! I'll see….
Jeanne:	I also need some Cinnamon mangoes.
Mary:	That's a good idea. Today they're sweet and juicy.
Jeanne:	Is it the same price?
Mary:	Nope! These are five for three dollars.
Jeanne:	Okay. No problem. I'll take twelve.
Narrator:	Freeze! How much does Jeanne owe for the 12 mangoes?
Mary:	Mmm! I'll see….
Jeanne:	That's all…Here's a crispy twenty-dollar bill…Keep the change.
Narrator:	Freeze! How much change has Jeanne left for Mary? Explain.

SOLO PRACTICE-20

Re-read the playlet *Mango Juice*, then answer the following questions.

1. Did Jeanne and Mary know each other prior to the encounter? Provide some textual evidence.

2. How does Mary advertise her mangoes?

3. Which mango brand is more expensive (per unit price)? How do you know?

4. How would you describe the temperament of each character in the Playlet?

5. Indicate the idiosyncratic response of the vendor.

6. Indicate the idiosyncratic intervention of the narrator.

7. Does Jeanne have enough money for the purchase? How do you know?

8. Make a list of all the adjective in the text, and indicate their meaning.

9. How would you characterize Jeanne?

10. Compose your own playlet to share with the class.

PROBLEM-21
PHONE SALE

GROUP TASK-21

In the 1990's, two telephone companies Flashy and Speedy were competing in New York. Flashy would charge 50 cents per hour in addition to a 3-dollar flat fee for service. Whereas, Speedy would charge one dollar per hour and no service fee.

1. For what quantity of hours would the total charge be the same, regardless of the company? You may use the table below or any other strategy of your own.
2. Which store would you patron? Why?
3. Write an equation for each potential transaction at Flashy and Speedy.

Phone Sale Table		
Hours	Total Charge - Flashy	Total Charge - Speedy
0		
1		
2		
3		
4		
5		
6		
7		
8		
9		
10		

SOLO PRACTICE-21

A. Write in algebraic form.

1. Number *n* increased by 7.

2. Number 8 decreased by *z*.

3. The sum of the variables *a* and *b* is twenty-four.

4. The difference between *m* and *p* is 0.

5. A quotient of *t* and *q*, decreased by the sum of *r* and *s*.

6. The square of *c* equals the sum of the square of *a* and the square of **b**.

7. The product of the sum of *a* and b, and the difference between *a* and b is fourteen.

B. Solve the following equations.

1). $10x = 56$

2). $3p - 12 = -18p + 93$

3). $a - b = 4$, and $a + b = 36$

4). $\frac{1}{2}m + \frac{1}{3}m + 17 = 20 - \frac{1}{6}m$

PROBLEM-22
METRO CARD

GROUP TASK-22

In February 2016, in New York City one single-ride MTA MetroCard cost $2.75. A Middle School student swiped the card only to go to school or a part-time job and return home. At the same time, the one-month MetroCard is available for $90.

1. Determine the discount, in dollars, a rider would benefit if the young strapnager purchased a MTA one-month fare in February 2016.
2. Determine the percent discount.

Note: The student used his card throughout the month, excepts on Saturdays and Sundays.

SOLO PRACTICE-22

Referring to to Problem-19, the Metropolitan Transit Autority plans to rise the fare in the following month (March 2016) by 6% for the single rides, while keeping the one-month fare constant. What would be the percent discount in March 2016 if the same rider utilized the one-month MetroCard.
Note: The rider would use his card twice a day and would stay home on Saturdays and Sundays.

PROBLEM-23
WATER FLOW

GROUP TASK-23

A cylindrical water container is one-half meter tall with a diameter of 1 foot. When the faucet is activated, the flow flows at a rate of 5.209 liters per minute. At this rate of water flow, determine the number of minutes it would take for the container to be emptied.

Note: 1 dm^3 = 1 liter
  1 foot = 30.48 centimeters
  π = 3.14

SOLO PRACTICE-23

huge cylindrical container is ¾ meter tall with a diameter of 2 foot. When the faucet is activated, it drops 2.188 liters per minute. At this rate, determine the number of hours (and minutes if any) it would take for the container to be completely emptied.

Note: 1 dm^3 = 1 liter
      1 foot = 30.48 centimeters
      π = 3.1416

ORGANIC-WAY MATHEMATICS: THE ESSENCE OF PROBLEM SOLVING
STUDENT WORKBOOK 5-6-7
Copyright © 2023

PROBLEM-24
SWEET LEMONADE

GROUP TASK -24

This morning, Mom made some lemonade that was found delicious by all children: Julienne, Shandra, Jude, Angelo, and Myriam. Mom's recipe is the following: 2 cups of water, 6 spoons of sugar, 2.5 lemons, and ¾ teaspoons of vanilla extract.
Complete the Sweet Lemonade Table.
What have you returned?

Sweet Lemonade Table				
People	Water	Sugar	Lemon	Vanilla
Mom	4 cups	6 spoons	2.5 lemons	$\frac{3}{4}$ T. spoon
Julienne	8 cups			
Shandra	2 cups			
Jude			7.5 lemons	
Angelo		30 lemons		
Myriam				$7\frac{1}{2}$ T. spoons

SOLO PRACTICE-24

Based on what you have learned from the previous activity, complete the table below.

People	\multicolumn{4}{c|}{Cake Recipe Table}			
	Flour	Sugar	Milk	Vanilla
Nordie	5 lbs.	1 cup	2.5 cans	$1\frac{1}{2}$ spoons
Guerda	10 lbs.			
Claude	22 lbs.			
Marly		9 cups		
Farah			28 cans	
Rose				$\frac{3}{4}$ spoon

PROBLEM-25
DIAMOND PARK

GROUP TASK -25

A private park (see sketch below) is to be renovated. The diagonals of the smaller rhombus measure 60 yards and 80 yards, whereas the diagonals of the larger rhombus are 180 yards and 240 yards.
Red roses are to be planted inside the smaller rhombus at $100.25 per yd^2, whereas green grass will grow in the rest of the park at $10.00 yd^2. Finally, the park will be secured will a fence at $5 per ft^2.
Determine the total cost of the park renovation.
Complete the table and the pie chart.
What have you observed?

SOLO PRACTICE-25

Construct a similar problem. Solve it in multiple ways. Share it with the class.

GLOSSARY OF CLUSTERED & RELATED TERMS

Question	What is asked to find / Objective of a task
Step	One of the many phases of a procedure or solution process
Pathway	Set of steps leading to a solution
Process	Strategy and skills used to solve a problem
Solution Process	Pathway toward the solution of a problem
Skill	Ability to perform tactical operations / Procedural knowledge
Strategy	Grand plan toward a problem solution
Analyze	Study-Break down -Break down into composing parts
Comment	Feedback on a production / Opinion of a reviewer
Investigate	Study to discover the truth discover the truth
Verify	Prove a stated fact, statement, a conjecture
Justify	To prove the accuracy or validity of a statement
Evaluate	Judge / Measure / Gauge / Opine on then value
Perform	Execute a task
Convert	Translate into a different form
Expression	Whatever is expressed
Pattern	Predictable repetition of events
Formula	General rule / Algebraic equation showing relationships
Inequality	Imbalance between two entities
Pythagorean Theorem	Mathematical law stating: "in any right triangle the square of the hypotenuse equals the sum of the squares of the other two sides
Blueprint	Sketch or design
Venn Diagram	Graphic organizer that compares two or more entities
Table	Graphic organizer with rows and columns
Frayer Model	Graphic organizer showing the characteristics of a concept
Anticipatory Guide	A mechanism to elicit prior knowledge via pre-set statements.
Math Riddles	Statements that personify mathematical concepts
Semantic Map	Graphic organizer seeking connections with the central concept
Circle Graph	Graphic organizer where the entire circle is taken as 100%
Graphic Organizer	Representation of a situation via table, flow chart, bar graph, etc.
Difference	Result of a subtraction / What makes things not the same
Sum	Result of an addition / Collection of all given elements / Total
Product	Result of multiplication
Quotient	Result of division
Remainder	Leftover from division / What remains after an equal distribution
Factor	Quantity that is multiplied. Example 3 or 4 in $3 \times 4 = 12$
Multiple	Quantity derived from a multiplication of numbers (factors)
Element:	Distinct entity of a set
Intersection	Subset of elements common to many given set
Inclusion	Subset of elements within a given set / Incorporation: Symbol $\subset$
Union	Set compiling all other sets, wherein elements is not repeated

GLOSSARY OF CLUSTERED & RELATED MATH TERMS
(continued)

Time Zone	Area on earth with the same time
Military Time	Time used by military personnel. Example, 1300 hour for 1:00 pm
Pacific Time	Standard time within the limits of the Pacific Time Zone
Eastern Time	Standard time within the limits of the Eastern Time Zone
Central Time	Standard time within the limits of the Central Time Zone
Mountain Time	Standard time within the limits of the Mountain Time Zone
Minute	One-sixtieth of an hour (time) or of a degree (angle)
Second	One-sixtieth of a minute (time) or of a minute (angle)
Hour	One-twenty-fourth of a day
Day	One rotation of the earth
Week	Sequence of 7 days from Monday to Sunday
Month	Sequence of 30 or 31 days; except February 28; but 29 in leap year
Semester	Sequence of 6 months
Trimester	Sequence of 3 months
Mph	Number of miles covered in one hour
Unit	Standard of measurement
Yard	Unit of linear measurement in the Customary System: 1 yd = 3 ft ≈ 1 meter
Foot	Unit of linear measurement in the Customary System: 1 foot = 12 inches
Inch	Unit of linear measurement in the Customary System: 1 inch = 1/12 foot
Meter	Unit of linear measurement in the Metric System: 1 meter = 1/40,000,000 of the Circumference of the Earth
Centimeter	Unit of linear measurement in the Metric System: 1 cm = 1/100 meter
Kilometer	Unit of linear measurement in the Metric System: 1 km = 1,000 meters
Millimeter	Unit of linear measurement in the Metric System: 1 mm = 1/1000 mete
Decameter	Unit of linear measurement in the Metric System: 1 dam = 10 meters
Hectometer	Unit of linear measurement in the Metric System: 1 hm = 100 meters
Square unit	Unit of area measurement of a plane figure: Generic term used when specific unit is unknown
Ounce	One sixteenth of a pound
Pound	Value of 16 ounces
Kilogram	Unit of weight or mass measurement in the Metric System 1 kg = 2.2 pounds

GLOSSARY OF CLUSTERED & RELATED MATH TERMS
(continued)

Fraction	A part of one whole. A subset of a group of things
Numerator	Number of parts considered from a total of equal parts
Denominator	Total number of equal parts after a equal grouping or separation
Ratio	Rapport between two entities – Comparison between two entities
Compare	Tell what is similar about distinct things; Also tell of the difference
Percent	Numeric value when comparing to 100
Decimal	Number that involves parts of a whole
Nearest	The closest to a given entity
Square Root	Number that is raised to the given second-degree power
Scientific Notation	Expression of a number using a decimal and a power of ten
Transformation	Repositioning of a geometric shape on a Cartesian Plane or Coordinate System
Translation	Transformation where a figure slides in a certain direction
Reflection	Transformation where a figure flips over a symmetrical line
Rotation	Transformation where a figure turns about a given point
Dilation	Transformation in size while preserving similarity
Data	Given facts, information
Average	Mean on a statistical data / Value for an equal distribution
Mean	Statistical element or average; Per capita value in equal distribution
Mode	Most frequent element of the data set; Most repeated element
Median	Element located in the middle of an ordered line of data elements
Range	Difference between highest number and lowest number of a data set
Pi (π) or 3.1416	Ratio of circumference to diameter of circle / Irrational number
Diameter	Special chord crossing the circle though the center. Diameter = 2 radii
Radius	Line segment joining the center of the circle and a point on the circumference
Circumference	Perimeter of circle = $\pi d = 2\pi r$ / Line that makes entire circle
Chord	Any line segment joining two points on the circumference of circle
Arc	Piece of the circumference
Sector	Part of the area of the circle limited by two radii
Compounding	Method of computing balance of a loan based on periodic renewal of the principal
Principal	Amount that is deposited, loaned, or borrowed
Interest	Profit earned by the principal
Rate	Rapport between the principal and the interest expressed in percent
Probability	Chance for an event to happen.
Experimental Probability	Chance of outcomes based on experiments or trials
Theoretical Probability	Chance of outcomes based on mathematical or scientific calculations

GLOSSARY OF CLUSTERED & RELATED MATH TERMS
(continued)

Even	Natural number divisible by 2
Randomly	Blindly, by hazard; Without looking
With Replacement	Probability where the number picked is put back into the set
Without Replacement	Probability where the number picked is not put back into the set
Angle	Zone of a plane figure limited by two rays
Supplementary Angles	Pair of angles with a sum of 180 degrees
Complementary Angles	Pair of angles with a sum of 90 degrees
Adjacent Angles	Pair of angles that are next to each other, with a common side
Vertical Angles	Pair of angles of opposite regions of a common vertex when two Vertical angles are congruent
Congruent Angles	Angles of equal measures
Reflex Angle	Angle that measures more than 180 degrees
Alternate Interior Angles	Angles on opposite sides of the transversal but inside of the parallel lines. Alternate interior angles are congruent
Alternate Exterior Angles	Angles located on opposite sides of the transversal but outside of parallel lines. Alternate exterior angles are congruent
Corresponding Angles	Angles on the same side of the transversal, but one inside and the other outside. Corresponding angles are congruent
Concave Polygon	Polygon where one angle "caves" into the shape; diagonal crosses outside shape
Convex Polygon	Polygon where all diagonal cross inside the polygon
Regular Polygon	Polygon with all congruent sides or all congruent angles
Degree	Unit measure of angles / Extent
Line Segment	Part of a line with 2 endpoints
Point	The smallest figure in Geometry / Dot
Altitude	One dimension of parallelogram being perpendicular to the base
Perimeter	Borderline of a plane figure
Side	Line segment limiting a polygon
Distance	Gap / line segment between two locations
Diagonal	Line segment joining two non-consecutive vertices of a polygon
Hypotenuse	Side opposite to the 90-degree angle in a right triangle The longest side, generally "c", proving that $c^2 = a^2 + b^2$
Triangle	Polygon with three angles or three sides
Scalene Triangle	Triangle with no congruent sides
Obtuse Triangle	Triangle where one angle measures more than 90 degrees
Acute Triangle	Triangle where all angles measure less than 90 degrees
Right triangle	Triangle with a 90° angle
Width	One dimension of a rectangle
Length	One dimension of a rectangle
Square	Rectangle with 4 congruent sides; Rhombus with right angles

GLOSSARY OF CLUSTERED & RELATED MATH TERMS
(continued)

Parallelogram	Quadrilateral with 2 pairs of parallel sides
Quadrilateral	Polygon with sides
Rhombus	Parallelogram with 4 congruent sides
Trapezoid	Quadrilateral with only one pair of parallel sides (USA & Canada)
Trapezium	Quadrilateral with only one pair of parallel sides (England)
Polygon	Space figures limited by line segments
Hexagon	Polygon with 6 sides or 6 angles
Pentagon	Polygon with 5 sides or 5 angles
Heptagon	Polygon with 7 sides or 7 angles
Octagon	Polygon with 8 sides or 8 angles
Nonagon	Polygon with 9 sides or 9 sides
Decagon	Polygon with 10 sides or 10 angles
Dodecagon	Polygon with 12 sides or 12 angles
Shaded Region	Darken area of a figure
Area	Number of square units within a 2-dimensional, plane figure
Volume	Number of cubic units in a 3-dimensional, space figure
Prism	Space figure whose surfaces or faces are polygons
Faces	Polygons that set the limits of prisms and pyramids
Rectangular Prism	Prism limited by 6 rectangles. Example: shoebox
Cylinder	Space figure whose two circular bases. Example: Roll of tissue paper
Lateral Area	Surface other than the bases. Surface on the side of the space figure
Surface Area	Total area of all the faces of a space figure
Height	Altitude of a space/plane figure. Line segment perpendicular to the base
Base	Line segment perpendicular to the altitude of a plane figure
Base Area	Area of a space perpendicular to the altitude of a space figure
Dimensions	Basic measures of a figure. Ex: Length and width of a rectangle
Monomial	Algebraic expression with one term
Binomial	Algebraic expression with two terms
Polynomial	Algebraic expression with many terms
Trinomial	Algebraic expression with three terms
Distributive Property	Multiplication of each term of one factor by those the other factor
FOIL Method	Distributive property using two binomial factors by multiplying the **F**irst terms of the factors, then the **O**uter terms of the factors, then the **I**nner terms of the factors and the **L**ast terms of the factors
Half-dollar	America coin equivalent to 50 cents
Dime	American coin equivalent to one-tenth of a dollar
Quarter	Coin equivalent to one-fourth of a dollar
Nickel	American coin equivalent to one-twentieth of a dollar
Penny	American coin equivalent to one-hundredth of a dollar
Variable	Entity that can change in an algebraic expression or equation.
Equation	Equality between two things. The equal sign (=) must be present

GLOSSARY OF CLUSTERED & RELATED MATH TERMS
(continued)

Function	Relation where each domain element (x) is associated with only one range element (y)
Algebraic Solution	Process of resolving a problem in algebraic terms using variables
Geometric Solution	Process of resolving a problem using geometric shapes
Graph	Graphic organizer showing relationship (function) between two or more entities
X-Axis	Horizontal axis in a Cartesian plane
Y-Axis	Vertical axis in a Cartesian plane
Coordinate Plane	Cartesian plane or grid / Coordinate system identified by two perpendicular lines or axes XX' and YY", intersecting at a O, origin of the axes
Slope	Steepness of a line / Constant and ratio between the coordinates of the points on the graph of a linear equation or function
Y-Intercept	Point of intersection between a graph and the Y-Axis
X-Intercept	Point of intersection between a graph and the X-Axis
Origin	Beginning of the axes / Intersection of the axes, re: Cartesian plane

ORGANIC-WAY MATHEMATICS: THE ESSENCE OF PROBLEM SOLVING
STUDENT WORKBOOK 5-6-7
Copyright © 2023

About the Author

Archangelo Joseph, Ed.D., is a New York-based education consultant. He provides consultancy services to the New York City Department of Education and to the private firm Generation Ready, LLC. Along with his wife Julienne, he has founded Organic-way Mathematics Consulting & Publishing, LLC. For 30 years, he has served public school children in multiple capacities: Bilingual Teacher, District Staff Developer, NYC Headquarters (Tweed) Resource Specialist, Math Coach, Assistant Principal, and New York State (Language RBERN) Resource Specialist. Among all of them, he prides himself to be a prolific curriculum and instruction designer, particularly in K-8 school mathematics and English as a New Language (ENL), formerly English as a Second Language (ESL).

In 2020, Dr. Joseph received his degree of Doctor of Education within the School Leadership Program at Concordia University Chicago. Espousing the constructivist epistemology of Lev S. Vygotsky and Jean Piaget, the doctoral candidate had conducted a 7-year research "A Case Study of Mathematics Teachers' Perceptions of Haitian Students Solving Constructed-Response Questions in New York Middle School" examining why students can't solve problems requiring justifications of answers. Before then, Archangelo had graduated from the College of Saint Rose, Albany, New York, with two Advanced-Study degrees (School Supervision and District Supervision). Equally important, he had mesmerized classmates, chiefly Evelyn Gerdes, at City College of The City University of New York City College (CUNY) with a Master of Science degree in Education with a (4.0 GPA).

Besides leveraging school leadership and pedagogy skills respectively among New York school decision-makers and teachers, Dr. Joseph along with his spouse and a few friends/colleagues traveled to his native Haiti in 2017 to establish the "Centre de Ressources Éducatives du Bas-Artibonite" (Center for Educational Resources of Bas-Artibonite). Over there, the team captivated educational stakeholders (300 teachers and 100 school principals) in the City of Saint-Marc and the City of Gonaives by the wisdom regarding best practices in school pedagogy and in school supervision and administration.

Dr. Joseph, a tri-lingual (English, French, and Haitian Creole), lives happily in Nyack, New York, with his other half Julienne and their youngest daughter Shayenne who, now, is completing her freshman year within the medical program at Hampton University in Virginia.

Made in the USA
Columbia, SC
19 July 2023